Quality by De

(Q

Volume

The Definitive Guide on Deviation

13 MAR 2025
TO CLARE'S SISTER
CHARLOTTE.
MICHAEL A. DELITALA

By: Michael A. Delitala

Dedicated to Troy Wilsey

Table of Contents:

Chapter 1

Meet Michael A. Delitala

Dear Reader, I'm extremely experienced in the pharmaceutical industry, with over three decades as mostly a "quality and compliance guy." This volume is dedicated to **Events** versus **Deviations** of which I am a subject matter expert (SME).

When it comes to Events & Deviations, I might just be one of the most experienced individuals in the business. That's not my ego speaking; that's high-level C-Suite leaders who took notice of my affection and affinity for the quality management system and my passion for these things we call *deviations*.

It was not always so. One of my very first gigs at Emergent BioSolutions had me as the quality reviewer and approver of deviations. As I was newly departed from the military, I seriously questioned my identity. What had I gotten myself into? Deviations are "horrible," and "I hate my life right now." As I pondered my existence, existentially, I started to grow very fond of deviations. In fact, for all of the quality management systems (QMS), deviations became my love. Most people think that anyone who loves deviations is a strange dude and perhaps a smidge of a sadomasochist.

At large pharma companies, I became known as Mr. Deviation, Mr. Hollywood, and Mr. Metric. All were badges of honor that I wore and still wear proudly today.

By the numbers, then, shall we?

At Emergent BioSolutions, we had a deviation process by which each Quality Approver had to author what we called the Final Quality Assessment (FQA). Most Quality Approvers would simply succinctly summarize the Owner's work. I used the FQA as an opportunity to complete my own investigation in parallel with the Owner. This way, I could get to know the business better. Also, many times, it allowed for an amazing conversation between myself and my non-quality partners. Especially if we arrived at a different conclusion which would occur on occasion.

In my 6 years there, I would say I was the FQA Author and Deviation Quality Approver on over 1000 total investigations. The deviations varied in subject matter and the ones I worked on included the full cycle of commercial drug substance and drug product manufacturing. And for a fermentation manufacturing process that was over 50 years old. This experience would go on to serve as my *education* for all future positions. And for that, I have one of the best bosses ever to thank for the opportunity. Mr. Troy Wilsey. Also, at Emergent, I was fortunate to work with one of the best Trainers I have ever known, Mrs. Sue Shoemaker. Sue taught me everything I know today about Training. And Sue let me author and execute a deviation training of which I won a guest-instructor award.

Following Emergent BioSolutions, I headed to Seal Beach for the first of its kind Sipeulecel-T, immunotherapy product at Dendreon. Their product was known at the time as Provenge, which is a spin on words that stands for *revenge on prostrate cancer*. At the time, the product had an 18-hour expiry. You read that correctly. I was in charge of the QMS. I worked the nightshift because Dendreon really needed all of my Emergent Experience. After all, on any given night, we were manufacturing 36 products. Pretend each product had at least one deviation on it. That meant we had to understand the product impact within that 18-hour window because there was a very ill, often times older man, waiting for his infusion to extend his life in some clinic while his product (his cells) were en route to him. Thirty-six deviations per night, times four nights, times two years is how many deviations I worked on at Dendreon. I hope I'm sketching a clear picture that I know my way around events and deviations.

These experiences paved the way for me to continue to be recognized for being a subject matter expert. At both Gilead (Kite) and Amgen, I was the program owner for deviations, and as you know, these names make up the top five in Pharma. I'm not sure that I'm allowed or permitted to say more regarding those two big names so head to my LinkedIn page to know more. In short, what I can offer is that I was tasked by both to reimagine the QMS, including deviations, and make it more user-friendly. So, that's what I did.

One last point. As the owner of the global programs (or just programs), you are the lucky individual to defend them with Regulatory Bodies, Partners, and Internally with Internal Auditors. I'm proud to write that my programs, other than at Emergent BioSolutions, where I obtained my education, not one time, did my programs receive any observations. Why is this noteworthy? It means my programs are compliant. This also means that if you're here for the answer, then the back of this book contains the most compliant Deviation SOP on the planet. I'm being serious. **Side Note:** For clarity with my former colleagues and former bosses, I did receive observations from an internal audit perspective, but they were pedantic at best and only caused more *busy* work to be completed. In other words, they did nothing to continuously improve the quality management system which is required by law.

I've been in the industry since 1995. As of this writing, that means I'm ancient and I have put in a dedicated twenty-nine years. For the majority of those years, in fact every single year, I've put heart and soul into the quality management system known as Deviation. And, I've trained greater than 5000 trainees on Deviations.

And in an attempt to succinctly summarize the whole of my experiences, I've worked pre-clinical bench scale, clinical phase 1, 2, and 3 in small molecule and biologics, clinical and commercial immunotherapy, and clinical (soon to be commercial) RNA Therapeutic. I've never worked in Medical Device, and I hope I never do. I love the inherent variability of biologics, and no offense to you Medical Device folks; it's just not for me. (CAPA before Deviation! Yuck! Micro iterations of change with intentionally planning to revert it back to the original state, again, Yuck!).

Chapter 2

What's the point?

By now, I hope you're wondering: Why am I reading this?

In short, I hope you have the opinion that I might actually be qualified to write about this subject.

I'm going to be very direct.

I wrote Volume 1 of QbD, **The Quality Manual** because I wanted to give the world the answer. I did not realizc until the work was published how therapeutic that writing it would be. Out here in California, I've had three bad bosses in a row. Just horrible, disgusting individuals who lack tact and emotional intelligence. You know, the kind of guy who thinks he can still raise his voice at you like he's your dad or something. So, I decided to take matters into my own hands and have formed the Delitala Quality Consulting and Author Corporation. I anticipate that there will be twenty total Quality by Delitala volumes. I mention this here because after writing Volume 1, I felt empty. Like how you feel after a very good session in therapy. So, my reasons are the same as before. I'll attempt to list them.

I want Quality employees who are new to the industry to have easy, digestible reference material so that it makes their understanding of what "quality" is supposed to mean expedited. For example, being exceptional with grammar is not "quality." That said, there are a plethora of resources that define "quality," so I'm not intending to "go there". Anyway, I want to give the answer away because conferences and consultants are expensive.

I'm tired of joining a company as a new employee thinking this will be the time I get to do this the way I want to do it. Inevitably, I report to some supervisor who wants it their way, and it's not the best that it can be, and I don't settle for less than the best. And my SOPs are simple, and simplicity breeds compliance. It usually becomes a problem because I'll provide excellent and then have it shredded to less than. (And, of course, you have to support the supervisor).

Also, start-up pharma companies or even established commercial pharma companies do not need to keep hiring expensive consulting firms to write the book of policies and procedures. The tome of policies and procedures has largely been in place since the 1950's. So let's not kid ourselves any longer big consulting firms!

CAPA will be the next volume!

Now comes my legal disclaimer:

My mind is a trap. Think of the guy on Suits who can remember the page and sentence placement for a specific word in a specific volume. My brain works somewhat like that (unfortunately). For example, I can remember the procedure number for the Good Documentation SOP from where I worked in 2004. So, for every company that I have worked for, please remember that you had me author a plethora of policies, standard operating procedures, work instructions, logs, forms, job aids, quick reference guides, training materials, investigations, CAPAs, EVs, Change Controls, and Quality Management Reviews. That said, what I author in this book is from my brain, my intellectual property, my experience, my invention, and the way I would write anything if I were still working for you or if I were the Head of Quality. Get it? It is 100% my work product. It is 100% my intellectual property.

Also, for any company that I am currently employed by or consulting for, I authored this outside of working or billing hours, generally at night during the week, on a vacation day, or the weekend.

Large consultant firms and perhaps even the industry personified itself will hate me for this.

And I'm okay with that.

Now, if you're just here for the answer, and you don't care about the "what" or the "why" or for Michael D to "wax poetic," then skip ahead to the appendices. There, in the appendices, I authored my version of the Deviation SOP and all other supporting controlled documents, including training!

Since you bought the book, you might require the answer. I hope your boss thinks it's an excellent work product (but they'll probably shred it!).

Good luck and enjoy.

You're welcome to connect with me on LinkedIn: www.linkedin.com/in/michaeldelitala

I intend to empty myself on all things related to Deviations.

Volume 3 will be dedicated to Corrective Action Preventive Action (CAPA).

Chapter 3
What's required?

Many other reference materials like this one will detail for you every single regulation from all regulatory bodies where the term Deviation is referenced. That certainly is an approach. My source of truth and my approach is to mostly pay attention to and reference the "Guidance for Industry Q10 Pharmaceutical Quality System".

Why: The Introduction of ICHQ10 explains that if you meet these guidance requirements, then you'll be compliant.

Two things here about what compliance means:

1. Compliance means following the regulations (the law).
2. Compliance also means meeting your own internal standards, which are the company's policies and standard operating procedures.

That's it. That's compliance so don't let any binary Head of Quality tell you anything different. If they do, send them to me. I'm being serious.

Please continue below for what ICHQ10 requires for Deviations.

ICHQ10 Process Performance and Product Quality Monitoring System:

Include feedback on product quality from both internal and external sources (e.g., complaints, product rejections, nonconformances, recalls, **deviations**, audits and regulatory inspections, and findings).

ICHQ10 Corrective Action and Preventive Action (CAPA) System:

The pharmaceutical company should have a system for implementing corrective actions and preventive actions resulting from the investigation of complaints, product rejections, nonconformances, recalls, **deviations**, audits, regulatory inspections and findings, and trends from process performance and product quality monitoring.

ICHQ10 Management Review of the Pharmaceutical Quality System:

Management should have a formal process for reviewing the pharmaceutical quality system on a periodic basis. The review should include:

a. Measurement of achievement of pharmaceutical quality system objectives.

b. Assessment of performance indicators that can be used to monitor the effectiveness of processes within the pharmaceutical quality system, such as:

1) Complaints **deviation,** CAPA, and change management processes.

2) Feedback on outsourced activities.

3) Self-assessment processes including risk assessments, trending, and audits.

4) External assessments such as regulatory inspections and findings and customer audits.

That's it. It seems a bit "light" to me for a quality management system that is known as a "heavy hitter." Please keep reading; I'll address why it's light in subsequent chapters.

What is this thing they keep referring to as the *pharmaceutical quality system*? It's defined in ICHQ10 as "The management system to direct and control a pharmaceutical company with regard to quality." What it means is that a company has to have a specific set of controls in place to ensure that reliably and consistently, a patient will receive a product that is safe, of the right potency, of the right efficacy, and that it has met all of its product quality attributes.

There are unique sets of controls put in place, they each have a quality system name, and per each quality system, it has an intended outcome. Deviations, then, make up one element of the Quality Management System (QMS). Refer to the below diagram by which I hope to illustrate this concept clearly. Hopefully, it's clear that the QMS is a principle to ensure a company, relative to quality is in a state of control. If more is needed, refer back to Volume 1

Figure 1: Diagram of the Quality Management System

Chapter 4

Quality by Delitala Tenets of Deviation Management

Tenet 1: All deviations are preventable.

Companies should deploy a culture through their values, mission statement, vision statement, goals, and objectives, detailing how to prevent deviations. There should be metrics. The metrics should be reported to Executive Management and Shareholders!

Employees should receive bonuses when it is determined that they have proactively prevented a deviation from occurring. The bonus should be commensurate to what it would have cost the employer.

For example, minor deviations cost employers >$1,000,000 annually.

Major deviations cost employers >$10,000,000 annually.

Critical deviations cost employers >$8,000,000 annually.

Imagine how rich an employee could be if they made it their mission to prevent deviations.

Imagine the profit for the company! (this is what Quality by Delitala is all about!)

Tenet 2: Each deviation is to be treated as a project.

Projects have milestones and due dates. So do deviations, but no one thinks like this. They're too busy worrying about a deviation due date.

Minor deviations, if they are truly minor, should be closed within 5 days from the initiation date.

Major and Critical deviations should be closed within 30 days, 100% of the time when each deviation is managed as a project.

Figure 2: Milestones for Minor Deviations

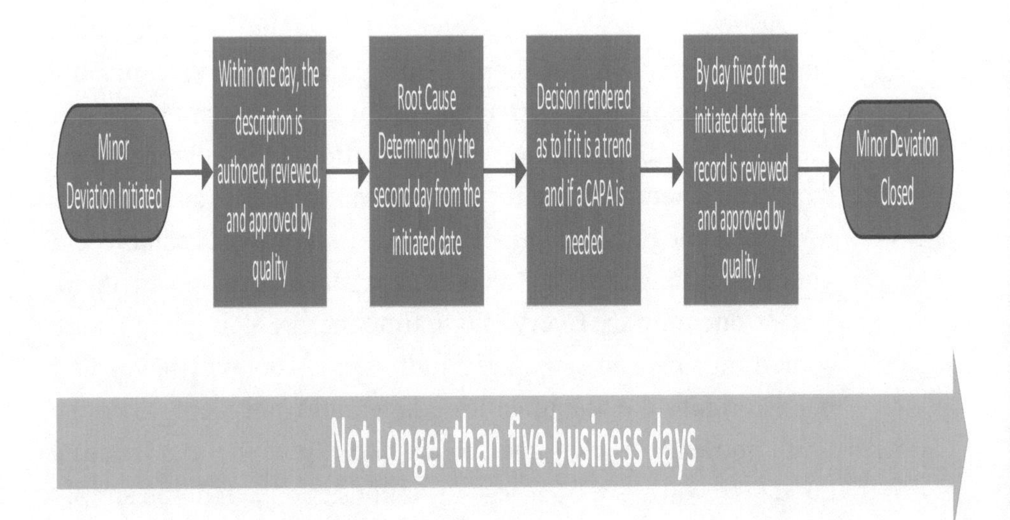

Figure 3: Milestones for Major/Critical Deviations

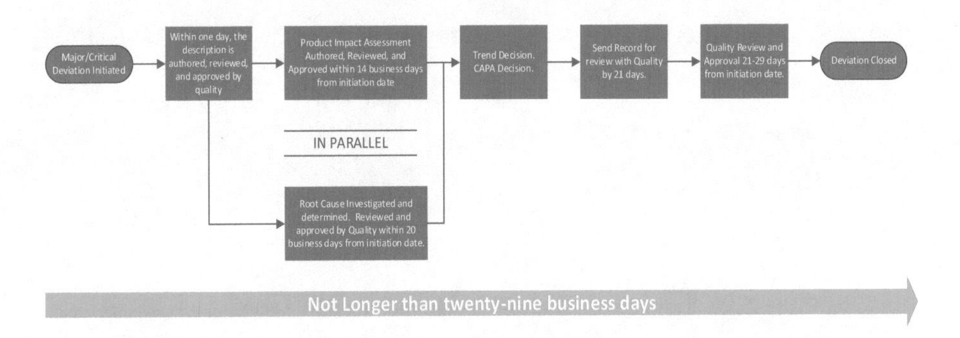

Tenet 3: Deviations shall not be viewed as problems.

Deviations are not problems. Therefore, deviations should never have "problem statements". Deviations have descriptions, not problem statements. Deviations are actually a company's greatest opportunity for continuous improvement. No one likes working on problems except me. So, to incentivize today's workforce, tell them about their opportunities and frame them in how you can save the company X dollars should they correct this from ever happening again. Place all of the deviations by deviation type into a Pareto diagram. Then, take the top three deviation types and evaluate for noticeable patterns. Since all deviations are preventable, get your best minds together to come up with ways to prevent them. Here's one on me: Every single time, before you manufacture something, take every single individual who will have their hands on into the manufacturing area and walk them down the line. Show them what good looks like. Have them practice. I guarantee you'll see 50% less deviations per batch if you simply walk your staff down the line and the process.

Tenet 4: Staff do not come to work to deviate;

Therefore, human error is never a root cause in minor deviations. If human error is cited, then it is automatically a major or critical deviation because advanced root cause analysis is required. See next tenet.

Tenet 5: Names are facts. Names are allowed in deviation records.

In nearly every company that I've worked for, Quality Leadership will not allow names to be placed in deviation records. The reason is because immature managers use deviations punitively. They punish their staff for deviating and for being cited as the root cause of a deviation. Stop that immediately! Traceability is a requirement, therefore, names and titles, as facts, are absolutely permitted in deviation records.

Tenet 6: If you don't have a centralized model for Deviation Ownership, then you need templates, lots of templates. And Field Entry Guides.

Deviations are not every staff's day job. Sometimes, staff might only work on a deviation once a year. Training is expensive. Therefore, have lots of templates so that people don't have to think about deviations, and then you can continue to pay them for what they're supposed to think about.

Once, a Regulator paid me a huge compliment by saying, "Gosh, you really don't let anyone think around here." Correct. It's why I have everything in my QMS standardized and templated. People have science to do over here!

Tenet 7: Thou shall not limit the number of pages for any standard operating procedure ever!

Heads of Quality and sometimes consultants have this strange notion about the number of pages in a standard operating procedure. It's really weird. And they can never tell you why they want this limit. It's usually ten pages. You're not allowed to have more than ten pages.

Here's a reminder about the document hierarchy:

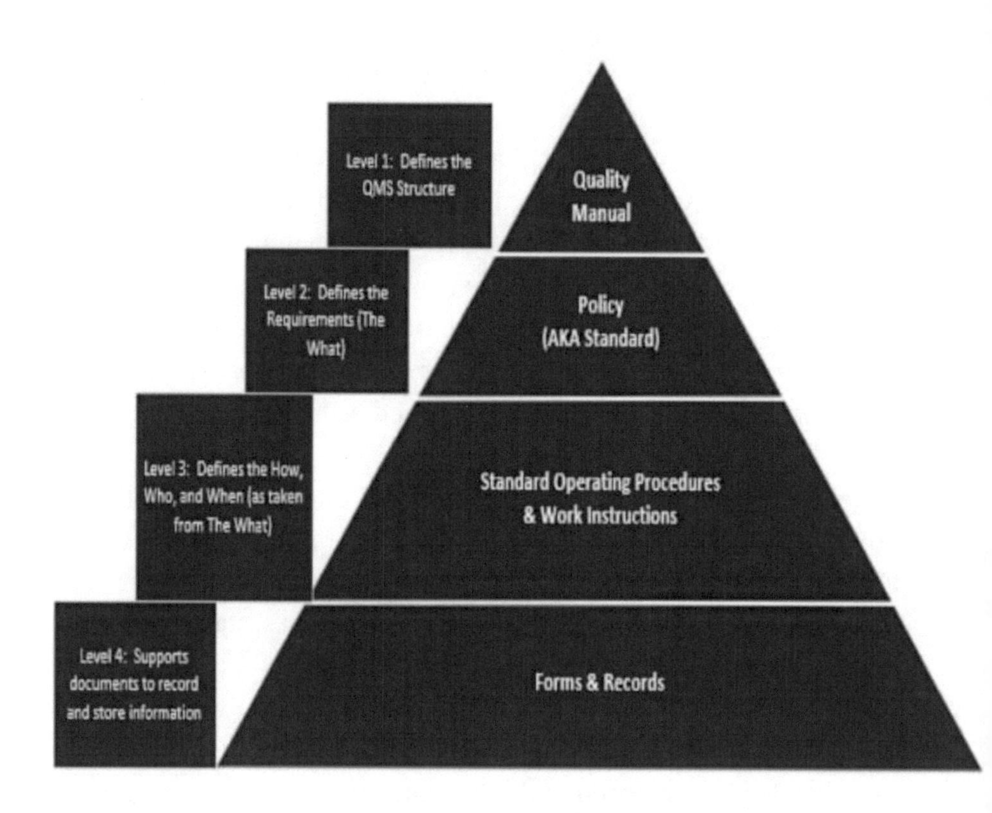

With this kind of hierarchy, they have a good point. If you define it correctly, you probably do not need more than 10 pages. The point is you can have as many pages as you'd like in your SOP. Regulators don't care about the number of pages; they care about the number of versions. Take, for example, the deviation management SOP. If your deviation management SOP is on version 30 and you're a three-year-old company, then it reflects on how out of control your company is. Seriously!

Revising the SOP thirty times in less than three years speaks volumes about your lack of quality leadership.
I recently had this discussion with a Head of Quality, and I even pulled out the FDA's SOP on SOPs. The FDA SOP on SOPs is eerily silent on the number of pages but fleshes everything else out in the detail that you'd expect. I pointed this out to him and he didn't get it. He only gave me more perpetual feedback, and he thinks he's right. Anyway…

Tenet 8: If you're an immunotherapy company, then Apheresis OOSs are not deviations. They are MRB records. Period!

Deviations get to an actionable root cause so that you can have an effective CAPA, and therefore, conceptually, that deviation will never occur again. Aph OOSs are the result of the inherent variability of a patient. They usually occur because a patient's cells are too weak to reproduce themselves in media. So, they are out of spec due to their viability. There is nothing anyone can do about that. So, it's a deviation, but it's not a deviation that can be corrected. Therefore, skip the deviation and go right to an MRB. For those that don't know, an Aph OOS can still be dispositioned and provided back to the patient. Actually, the FDA expects this because to take someone's cells and not give them back causes more harm than good.

Tenet 9: Deviations can have more than one root cause! They can have as many as the Owner identifies. See the QbD Volume on Root Cause for more.

Tenet 10: All aspects of the QMS require a process flow diagram!

The level of maturity of the QMS can be identified in each SOP, especially if the SOP does not contain a process flow diagram. For each QMS element within the SOP, there must be a process flow diagram. The process flow diagram is what should be used to defend the QMS to a Regulatory Authority.

Each element of the QMS is a process, and a process has outputs. In the Deviation Management Space, the outputs of the process are the following:

1. A Description
2. Product Impact Assessment
3. Impact Assessment
4. Root Cause(s)

` Here is the Quality by Delitala process for Deviations:

Figure 4: High-Level Event to Deviation

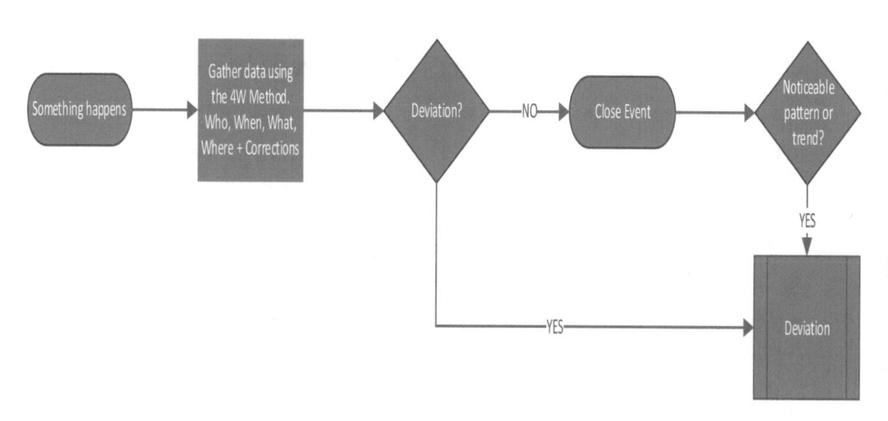

Figure 5: High-Level Deviation Process Flow

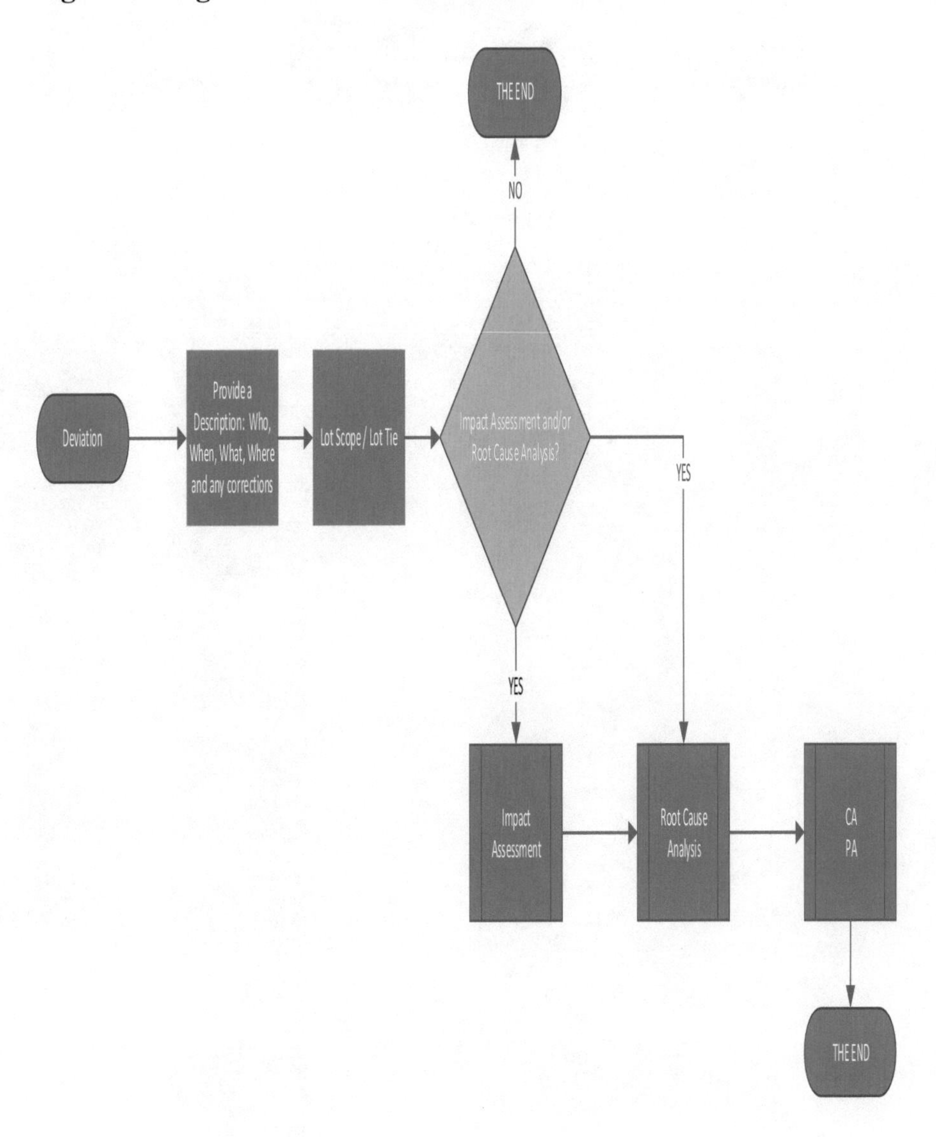

Figure 6: Detailed Level Deviation Process Flow

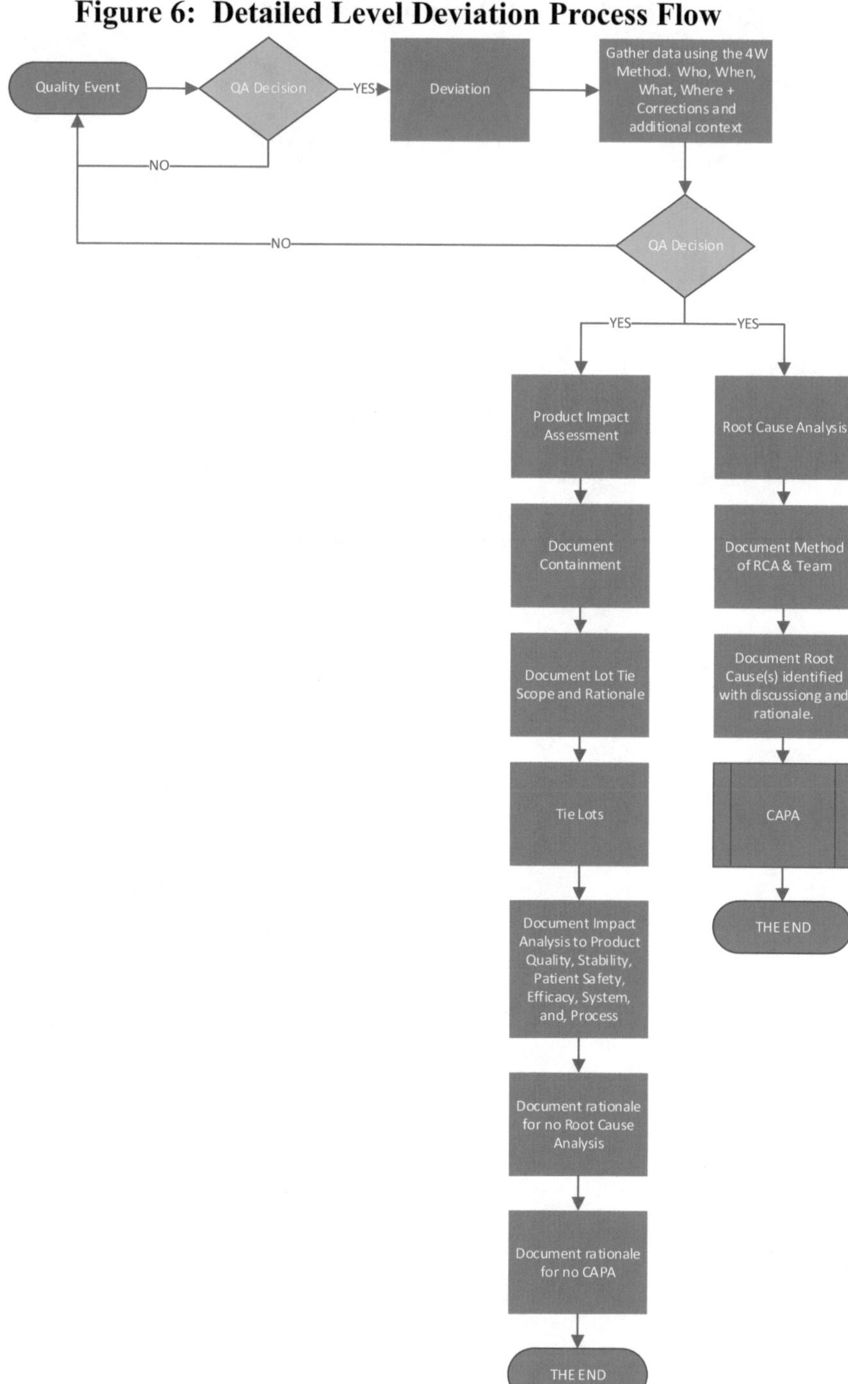

Tenet 11: Your company is not having a good inspection if there are a lot of deviations being presented to Regulators. Your company is having a good inspection if only executive summaries of deviations are presented.

Tenet 12: Only tie lots that are in the scope of the deviation. The business of lot tying is rampant in the industry. It takes Deviation Owners more time to tie lots than any other aspect of a deviation record. The systems we use for deviation management are not lot tie friendly. They should be! Trackwise requires a grid and a manual entry. Veeva uses an object and an ingestion of data, which is also manual. It takes Deviation Owners a lot of time to insert the lot numbers. It takes Quality even more time to verify the numbers. This costs pharma companies a lot of money. It doesn't have to.

First of all, conservative Heads of Quality require that all potential lots that are in the scope of the deviation are tied. This is not required, and second, I would argue it actually poses a greater risk to your state of compliance than if you only tie the lots that are necessary to be tied.

Here is a proposal for your rigid boss to consider:

Proposed Definitions to the Revised Lot Tie Process

Term	Definition
Distributed Product	Product that is outside of company control.
Product Impact	An event that results in impacted product/material being rejected in the inventory control system when it is factually determined there is an adverse impact on product quality.
Tied Lots	• Lot/batch where the deviation occurred • Lot/batch where product impact assessment is unknown, requiring additional data to determine impact and potential inventory control.
Routine Tests	A standard catalogue of procedures or tests that are regularly performed as part of the normal manufacturing process.

Proposed Revisions to the Standard Operating Procedure

Deviation Initiation

- In the lot tying field/grid, based on the definition of Tied Lots, include lot numbers of products and/or materials tied to the event.
- In the lot tying strategy field, which includes Scoping Strategy, document the initial rationale for the lots tied, or rationale for no lots tied, and/or lots considered but not tied.

Finalize risk categorization, scoping strategy, and tied lots

Scoping Strategy

- The Deviation Owner documents the scoping strategy in the Lot Tie Strategy Field and includes the following:

IF the lot-tied grid has…	THEN document in lot tie strategy the rationale for…
No tied lots	No tied lots
Tied lots	Tied lots

- The rationale for lots considered under the investigation but not tied (in scope) including but not limited to the following:

IF the deviation is related to..	THEN scoping strategy must…
LIR (Analytical test results)	Consider all lots produced between the last acceptable test result or when the last acceptable requirements were met prior to the deviating event AND the next acceptable test result or requirements were met after the deviating event.
Raw Material OOS during Full company specification (FCS) testing profile	Include all previous lots tested per company abbreviated specification (CAS) profile since last passing FCS.

- Document rationale for lots considered (in scope) but not tied based on the following:

IF scoping strategy rationale is…	**THEN ensure Lot Tie Strategy includes…**
Reproducible from a GxP source	- GxP Source for query - Start/End Date of query - Who performed the query - When the query was executed - The rationale for lots considered and not tied
Not reproducible from a GxP source	- The rationale for lots considered and not tied. - Attach supporting data to the deviation record.

Tenet 13: If you have any clinical products in your pipeline then all deviations must assess impact to the clinical trial application (CTA). If you have any commercial products in your pipeline, then all deviations must assess the impact to the marketing authorization (MA).

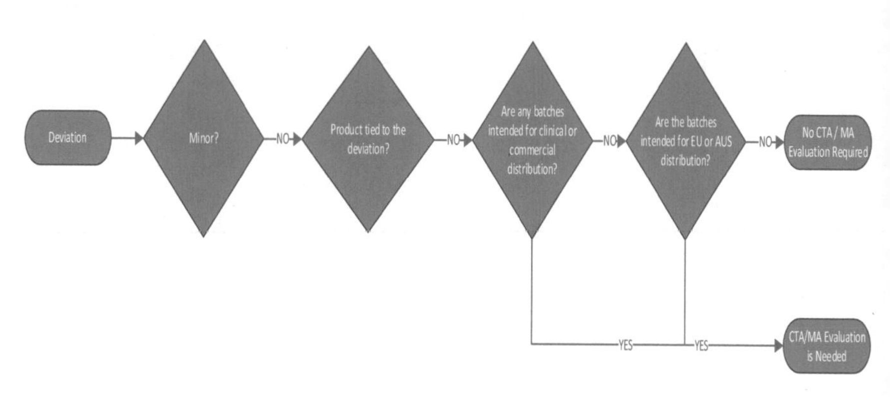

Tenet 14: The full of the QMS must have a contingency plan regarding when the computerized system is down. (And no, I'm not giving that to you!).

Appendix 1

The Quality by Delitala Deviation Standard Operating Procedure

Title: Deviation Management

Table of Contents

1. Purpose

1.1. The purpose of this standard operating procedure (SOP) is to define the deviation management process, which includes observation, notification, risk determination, investigation, impact assessment, root cause determination, corrections, evaluating for corrective actions, documentation, and reporting of deviations in the quality management system.

2. Scope

2.1. This procedure is required for all GxP activities for all GxP staff in all GxP departments at Callirrhoe Biosciences (Callirrhoe).

2.2. Validation deviations are not in scope of this procedure as Validation deviations are denoted as Exceptional Conditions per SOP-12345, "Management of Validation Deviations".

3. Responsibilities

Role	Responsibilities
Every Employee	• Following procedures while performing any cGxP operations or computer system activity. All employees must notify their management in the event they notice any departure from any required written instructions.
Authorized Person	• Provide continuing support for deviations impacting the local country's marketing authorization.
Deviation Initiator	• Observe (Identify) and initiate a deviation. • Determine if a correction is necessary, implement it, and document it. • Notify area supervisor and quality personnel of the deviation. • Collects initial facts and data on the event and provides it to the deviation initiator. • Categorizes the deviation and documents the rationale for the categorization.

Role	Responsibilities
Deviation Owner	• Manage the Deviation through specific milestones as detailed in the procedure. • Performs the investigation, including the root cause analysis. • Complete or assign tasks to complete the investigation timely. • Complete the assessment of risk to determine the deviation classification. • Gathers subject matter expertise for the investigation when needed.
Department / Area Supervisor / Manager	• Supports the investigation, when needed by Owners in their department. • Review the deviation details for completeness, correctness, and concurrence as required and provide an e-signature for Major and Critical deviations. • Monitors deviation for adherence to due dates and reallocates resources towards the timely closure of investigations as applicable or as needed. • Ensures deviation investigations within their departments are robust, accurate, and complete and that deviation records are reviewed and evaluated prior to the Quality review and approval. Provides feedback to deviation owners.
Process Development (Product Sciences)	• The authority and decider of the impact to product. • Authors of product impact assessments whenever product is tied to a major or critical deviation.
Supply Chain	• Executes inventory status reports for tracking and tracing of the disposition and distribution status of lots tied to major or critical deviations.

Role	Responsibilities
Quality Assurance (Denoted as Quality)	• Review deviations for completeness, correctness, and concurrence prior to closure. • Review deviation classification and decide which class the deviation is. • Manage deviation workload to ensure investigations are closed timely. • Ensure appropriate inventory status relevant to specific conditions documented within a deviation record.
Qualified Person (QP)	• Reviews and approves the impact assessment as an additional approver for major and critical deviations associated to lots designated for Europe.
Regulatory Affairs	• Assess deviations identified post-production, if distributed, for the potential to report a BPDR, per SOP-23456, “Biologic Product Deviation Report”.
Responsible Person	• Provide continuing support through the lifecycle of all deviations involving product distributed within a country, or a process related to a country.

4. References

4.1. Add References that are contained in the body of the SOP.

5. Definitions

Term	Definition
Affected Materials	Products, parts, and materials that are either nonconforming or are associated with a specific deviation and have the possibility of being nonconforming.
Biological Product Deviation Report (BPDR)	A required report submitted to the FDA for licensed distributed biological products for which a deviation was identified post-distribution – Refer to SOP-78901.
CAPA	Corrective Action Preventive Action – Refer to SOP-34567. A systematic approach that includes actions taken to eliminate the confirmed root cause of a detected deviation in order to prevent recurrence (Corrective Action). Preventive Actions (PA) are additional actions completed to eliminate the cause of a potential deviation in order to prevent occurrence. Preventive Actions also address contributing factors related to a deviating event, in addition to, or in lieu of a definitive root cause.
CMO	Contract Manufacturing Organization. A company contracted to perform GMP activities on behalf of Callirrhoe. Refer to SOP-45678.
Collaboration	The act of providing input to the draft version of a deviation document in the form of electronic comments, questions, and/or editing of an electronic record.

Term	Definition
Containment (Mode of Correction)	Immediate separation / isolation of the identified problem to prevent it from being processed further or mixed with conforming product.
Correction	An immediate action taken to correct a deviated without addressing the root cause of the deviation.
Contributing Cause	Any factor that impacts the failure whose elimination will not prevent recurrence of the failure. It indirectly affected the outcome or occurrence. On its own, the cause might not have sufficient power to result in the event taking place.
Critical Deviation	A deviation with confirmed, or high likelihood of significant impact to product quality, safety, efficacy, data integrity, systems, and/or regulatory filing. Impact to Critical Processing Parameters (CPPs), Critical Quality Attributes (CQAs) and/or product impact is evident or determination of product impact requires additional non-routine tests or extensive analysis of existing data sets.
Critical Process Parameter (CPP)	A process parameter whose variability has an impact on a critical quality attribute and therefore should be monitored or controlled to ensure the process produces the desired quality.

Term	Definition
Critical Quality Attribute (CQA)	A physical, chemical, biological, microbiological property or characteristic that should be within an appropriate limit, range, or distribution to ensure the desired product quality. Product attribute related to safety and/or efficacy of the product.
Deviation	Departure from a written and approved procedure or other established standard for GxP operation and/or an unexpected event or failure.
Deviation Risk Category	Risk category applied to each deviation based on the impact of the deviating event.
Deviation Type	Broad classifications of deviations used for trending purposes and to provide guidance on which departments are to be involved in the investigation. Only one deviation type may be selected per deviation.
Disposition	Planned action to be taken on product to correct, accept, or reject.
Distributed Biological Product	**Immunotherapy Definition:** • A licensed Callirrhoe product that has been released for infusion and delivered to an infusion site.
Effectiveness Verification (EV)	The process utilized to determine if an associated action was effective at eliminating and/or preventing the root cause of a specific deviation.

Term	Definition
eQMS	Electronic Quality Management System: Computerized Quality Application for documenting, tracking, and managing deviations and other quality systems elements and records (e.g., CAPA, Change Control etc).
Event	An occurrence during GxP operations that has the potential of being a deviation from cGMP, regulatory submissions, validated parameters, standards, policies, procedures, or specifications. All events are documented. Documentation of the event should be to the extent necessary to address potential product, process or facility impact.
External Deviation	Deviations tracked at Callirrhoe that occurred at and are owned by external CMOs.
Failure Mode	Points of failure in a process or system resulting in a deviation. Failure modes are divided into categories and subcategories. Failure modes are often known as the 6Ms.

Term	Definition
Functional Area Manager	Functional Area Manager (or designee) of the department where the deviation occurred or where the CAPA or action items have impact or where they will be owned.
Immediate Correction	An immediate action taken to eliminate or mitigate the effects of an existing deviation, defect, or other condition or situation that is outside of the normal process.
Impact Assessment (IA)	A summary of the analysis of product (i.e. Safety, Identity, Strength, Purity, Quality), system, process, or other quality attributes that may have been impacted by the event. This includes data evaluated and the justification for the impact conclusion.
Impacted Lots	Product and/or material lots that meet one or both of the following criteria: • Product and/or material lots directly related to the event (example: the lot being manufactured or tested when the event occurred). • Product and/or material lots manufactured, stored, or tested within the timeframe of the deviating event and determined to be in scope.
Impacted Site	The site in scope, or impacted by the deviating event. If the deviation occurred at site, then the site is in scope and is impacted.

Term	Definition
Initiating Site	The site where the deviation was observed is the Initiating Site.
Internal Deviation	Deviation that is initiated to address a process performed within Callirrhoe's operations. Example: Manufacturing processes, internal laboratory processes, etc.
Investigation	The process of using inquiries and examinations to gather facts and information to address a deviating event or to resolve an issue to ensure the root cause is addressed and, if appropriate, CAPA are determined, assigned, and evaluated.

Term	Definition
Major Deviation	Deviation with potential moderate impact to product quality, safety, efficacy, data integrity, systems, and/or regulatory filing. Potential impact to CPPs or CQAs. Determination of Product Impact requires an investigation and analysis of routine testing or existing data. Additional routine tests may be required.
Material Review Board (MRB)	Senior Leadership from Quality, Product Sciences, Manufacturing, Regulatory Affairs, Medical Affairs, and Clinical Development who provide subject matter expertise on the matter of final product failing to meet established acceptance criteria, or on matters related to the potential of final product failing any quality attribute.
Minor Deviation	Deviation with no potential impact or negligible risk to product quality, process, safety, efficacy, data integrity, and/or systems. No impact to CPPs or CQAs.
Objective Evidence	Qualitative or quantitative information, records, statements of fact based on eye witness account, measure, or testing which can be verified.
Owning Site	The site that owns the deviation and will complete the required actions to close it.

Term	Definition
Out of Specification (OOS)	A test result from a valid assay that fails to comply with the acceptance criteria (descriptive or numerical) given in an approved protocol or specification established by the manufacturer or official compendia.
Probable Root Cause	A factor, or a combination of factors that cause a deviation event but cannot be proven to be the root cause.
Process Comment	A comment captured within GxP documentation that describes an event that occurred that was not categorized as a deviation, but requires documentation. Events are simple and do not require root cause analysis or additional actions outside of controlled documents.
Recurring Deviation	An event that occurred again after implementing a CAPA to address the same root cause of the related event.
Repeat Deviation	An event that occurred again prior to or during the implementation of a CAPA to address the same root cause of the related event.
Root Cause	The underlying reason for the deviation.
Root Cause Analysis (RCA)	TBD in a later Volume of QbD.

Term	Definition
Subject Matter Expert (SME)	An individual possessing in-depth knowledge of a particular subject based on their education, training, and experience.
SQuIPP	An acronym for attributes used to describe the suitability of products: **S**afety, **Q**uality, **I**dentity, **P**urity, **P**otency
Retrospective Review	A review of the events that occurred during a defined time period to identify repeat or recurring deviating events and trends.
Trend	TBD in a later Volume of QbD.
Workflow	Workflow is the process of electronically circulating documents for review and approval to specific individuals. The workflow process utilizes electronic signatures for document approval.

6. Equipment & Materials – N/A

7. Safety – N/A

8. Procedure

8.1. General Considerations

8.1.1. Refer to Attachment 1 for the Deviation Management Process Flow Diagram.

8.1.2. This procedure shall be followed to identify and report unexpected events that occur and deviate from a standard or requirement to assure evaluation of impact, determine the need for further investigation into root cause, and assure nonconforming product is appropriately managed and dispositioned.

8.1.3. Refer to the Deviation Management process flow in Attachment 1.

8.1.4. Timely initiation of a deviation is essential in ensuring no impact to patient safety or product quality. It is the best practice to meet with Quality Assurance to decide if an event is a deviation and this decision is intended to be made within 24 hours of first becoming aware of the event. If the deviation is initiated after the first 24 hours, then a note documenting the reason for delay shall be entered into the Chronology of Events section.

8.1.4.1. Documentation Errors

8.1.4.1.1. Simple documentation errors that can be corrected following standard GMP documentation practices are not deviations. They are Process Comments per SOP-56789.

8.1.4.1.2. Documentation errors that can be addressed with a GMP source document or another GMP system/procedure are Process Comments per SOP-56789.

8.1.4.1.3. Documentation errors such as calculation errors, entry errors, transcription errors that cause results to be misreported and that have the potential to impact a material, product, or process will be classified as major or critical deviations.

8.1.4.2. Notifications

8.1.4.2.1. Initiator or Owner will notify area management immediately of any deviating event. When the deviation is initiated, the notification will be documented in the Chronology of Events with Who, What, Where, When, and How Much or How Many when applicable.

8.1.4.2.2. Initiator or Owner or Functional Area Manager, will notify Quality Assurance and all departments or sites that may be affected or impacted. The notification will be documented in the Chronology of Events with Who, What, Where, When, and How Much or How Many when applicable.

8.1.4.2.3. Quality Assurance will notify Regulatory Affairs regarding any deviation that may affect a released and delivered licensed Callirrhoe product.

8.1.4.3. Notify Executive Quality Management for any of the following:

8.1.4.3.1. Media simulation failures

8.1.4.3.2. Major supply disruptions

8.1.4.3.3. Atypical and/or significant impact to SQuIPP

8.1.4.4. Affected Materials

8.1.4.4.1. Quality Assurance will segregate any material affected by a deviation as deemed by Quality Assurance Management. QA will contact Supply Chain-Operations if the material requires special storage conditions.

8.1.4.4.2. If a MRB meeting occurs at part of a deviation investigation, the meeting is to be held per SOP-67890 and the meeting minutes are to be attached to the deviation investigation as part of the deviation closure.

8.1.4.5. Post-Release Distribution Deviation Events

8.1.4.5.1. Quality Assurance will notify Regulatory Affairs within 24 hours of the discovery of the deviation event that is identified post-product disposition & distribution to ensure that the event is evaluated in a timely manner for possible submission to the FDA as a BPDR per SOP-78901.

8.1.4.6. Environmental Monitoring Excursions

8.1.4.6.1. Refer to SOP-89012 for documenting investigations related to EM excursions.

8.1.4.7. Supplier Quality Deviations

8.1.4.7.1. If the deviation is associated with a Supplier, then notify Supplier Quality. Refer to SOP-98765.

8.1.4.8. Microbial Contamination (Sterility)

8.1.4.8.1. Refer to SOP-90123 for documenting investigations regarding sterility.

8.1.4.9. Yield – Out of Range

8.1.4.9.1. Refer to SOP-01234 for documenting investigations on Yield outside of set range.

8.1.4.10. Trend Investigations (Refer to a future volume of QbD)

8.1.4.11. Due Dates & Extensions

8.1.4.11.1. The due date for deviations is thirty calendar days from the date of initiation.

8.1.4.11.1.1. Due Date Extensions

8.1.4.11.1.1.1. Minor deviations are prohibited from being extended.

8.1.4.11.1.1.2. Major and critical deviations may have their due date extended only when the functional area department head and the head of quality agree to the extension inclusive of the next due date.

8.1.4.11.1.1.3. Past due deviations of all classifications require an update to the chronology of events section every week that the deviation remains past due. The intent of this requirement is to ensure that the deviation has timely updates as to the status on what is outstanding or incomplete in order to be able to close the record.

8.1.4.11.1.2. Investigation Progress Updates

8.1.4.11.1.2.1. In the event that a deviation investigation will extend beyond thirty days, the Deviation Owner must provide Quality with a documented progress update indicating the reason for extending the investigation, any associated risks and risk mitigations and a detailed action plan including expected completion dates for the remaining activities. Investigation Progress Updates must be signed the Owner, Management of the Owner, and Quality.

8.1.4.11.1.2.2. Additional updates are to be provided by the Deviation Owner on a monthly basis until the investigation is complete (the deviation record is closed).

8.2. Deviation Management Process – Deviation Initiator / Deviation Owner

8.2.1. Observe an event and evaluate if the event is a deviation. Refer to Attachment 1 for the decisions and steps in order to complete the evaluation in a consistent approach.

8.2.2. When Quality Assurance decides the event qualifies as a deviation, then the Deviation Initiator will initiate the record in the eQMS.

8.2.3. Determine if a deviation should be escalated to management. Refer to Attachment 2 and complete the actions as required by Attachment 2.

8.2.4. Determine the risk based approach for the deviating event and the required notifications. Refer to Attachment 3 to complete the steps.

8.2.5. Deviation Initiator / Owner: Describe the Deviation Event

8.2.5.1. The description of the deviation will contain, What, Where, When, Who, and when applicable How Much or How Many. The description will contain what was expected to occur and what actually occurred. Refer to Attachment 3 for examples of Descriptions.

8.2.6. Deviation Owner: Link Batches to the Deviation Record

8.2.6.1. The deviation owner will determine if the deviation requires batches to be tied to the deviation record. This ensures that batches will not be dispositioned nor distributed while the deviation record is in process.

8.2.6.2. The Quality Assurance approver will confirm the correct lot tie and document their evaluation in the Chronology of Events section.

8.2.7. Process Development: Determine the Impact to Product

8.2.7.1. Determine if the deviation event has actual or unknown product impact.

8.2.7.2. If an evaluation is required to determine the product impact, then the deviation will be classified as major or critical.

8.2.7.3. If an evaluation determines there is no product impact, then classify the deviation as minor.

8.2.7.4. Quality Assurance will confirm the correct classification. The confirmation is documented in the eQMS by an e-signature promoting the deviation record to the next step of the record lifecycle (Initiation to Impact Assessment to Root Cause Analysis).

8.2.8. Deviation Processing by Quality Assurance

8.2.8.1. Review the initiation information for completeness, clarity, and accuracy.

8.2.8.2. Quality Assurance will partner with Deviation Initiators and Owners and in real-time provide feedback and recommended edits to improve the quality of the record.

8.2.8.3. Quality will review the audit trail of the deviation record to ensure that the deviation is not a duplicate of an existing deviation. Duplicate deviations will be made void by Quality.

8.2.8.4. Quality will evaluate the extent / scope of the deviation and ensure no additional actions are necessary such as tying additional lots.

8.2.8.5. Quality will notify QA Operations if segregation of materials is warranted. QA Operations will follow QA Operations SOPs for relevant segregation steps.

8.2.9. **Deviation Owner: Complete the Impact Assessment**

8.2.9.1. Impact Assessments are different than Product Impact Assessments in that the intent of the Impact Assessment is to further evaluate the deviation for impact to system, process, data integrity etc…

8.2.9.2. All deviations require an impact assessment regardless of classification.

8.2.9.3. Deviation Owners will meet with subject matter experts based on the scope of the investigation to determine impact.

8.2.9.4. Appropriate and approved resources can be used as guidance to evaluate the outcome of the impact assessment. The sources must be verified and approved as a controlled document via SOP-87654, "Document Control".

8.2.9.5. The completed impact assessment must address the impact to the product, people, material, equipment, system, data integrity, and/or process.

8.2.9.6. If an MRB was held during the impact assessment phase of the deviation record, then the a copy of the meeting minutes must be attached.

8.2.10. Deviation Owner: Determine the Risk Categorization

8.2.10.1. Using the definitions for minor, major, and critical, evaluate the facts of the deviating event and conclude the risk categorization. Document the justification within the deviation record.

8.2.10.2. Quality Assurance is the authority and decider on the final risk categorization. Their e-signature on the record confirms the final risk categorization.

8.2.11. Deviation Owner: Perform and Document the Root Cause Investigation (RCI)

8.2.11.1. If a Root Cause Investigation is required, then the deviation record cannot close until the RCI is completed and approved by Quality.

8.2.11.2. The planning and execution of a Root Cause Investigation is managed via SOP-12859 (Note: Root Cause Analysis will be an extensive future QbD Volume).

8.2.12. Deviation Owner: Perform and Document the Retrospective Review

8.2.12.1. The retrospective review is completed from the initiation date to determine if the deviating event is new, repeat, or recurring. The review will include the duration of 2 years retrospectively from the initiation date.

8.2.12.2. Document the criteria used to complete the retrospective review.

The evaluation may lead to noticeable patterns in the data which may be viewed as a trend. Evaluate the need for any trend investigations, or CAPAs based on the results.

8.2.13. Deviation Owner / CAPA Owner: Determine CA & PA

8.2.13.1. Corrective Actions and Preventive Actions should be evaluated and determined based on the deviation classification. If CA / PA is to be initiated, refer to SOP-76543.

8.2.14. Deviation/CAPA Owner: Finalize and Route Deviation Record for Closure

8.2.14.1. The Deviation / CAPA Owner reviews the deviation record in its entirety for completeness, correctness, and accuracy and provides an e-signature which promotes the record to Quality.

8.2.14.2. Quality then reviews the deviation record in its entirety for completeness, correctness, and accuracy and provided an e-signature which promotes the record to closure.

8.3. Cancellation of Deviations

8.3.1. Deviations may be cancelled at any time by Quality.

8.3.2. The reason for cancellation is documented within the audit trail of the deviation record.

8.4. Reopening of Deviation Records

8.4.1. Deviations may be reopened at the discretion of quality to correct an error.

8.5. External Deviations

8.5.1. Callirrhoe tracks external deviations occurring at CMOs or Apheresis sites (Immunotherapy Only)

8.5.2. CMOs or Apheresis sites notify Callirrhoe of deviating events in a risk based approach. Only Major and Critical events are reviewed by Callirrhoe. This is regulated by Callirrhoe through NDAs, Service Contracts, and Quality Agreements per SOP-65432.

9. Attachments

9.1. Attachment 1: Deviation Management Process Flow Diagram

9.2. Attachment 2: Deviation Decision Matrix

9.3. Attachment 2: Management Escalation Decision Matrix

9.4. Attachment 3: Risk Based Approach to Managing the Deviating Event

Attachment 1: Deviation Management Process Flow Diagram

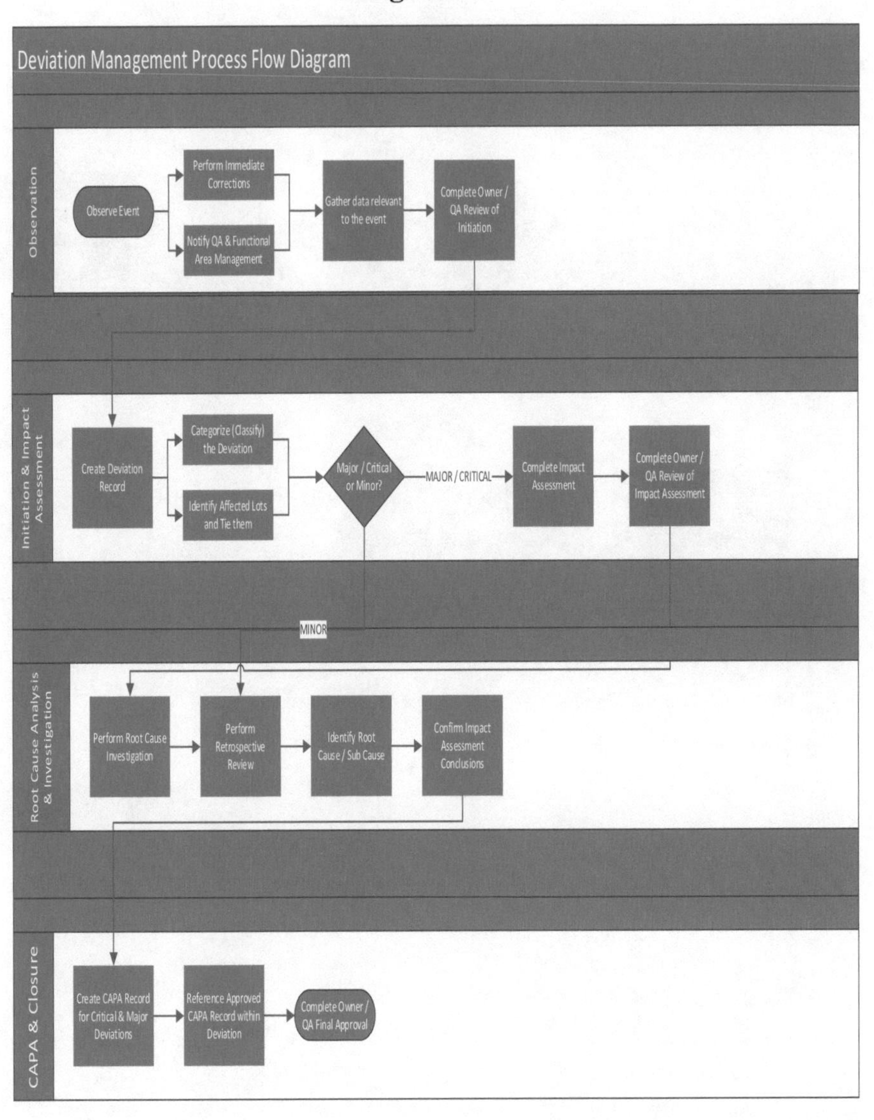

Attachment 2: Event versus Deviation Decision Matrix

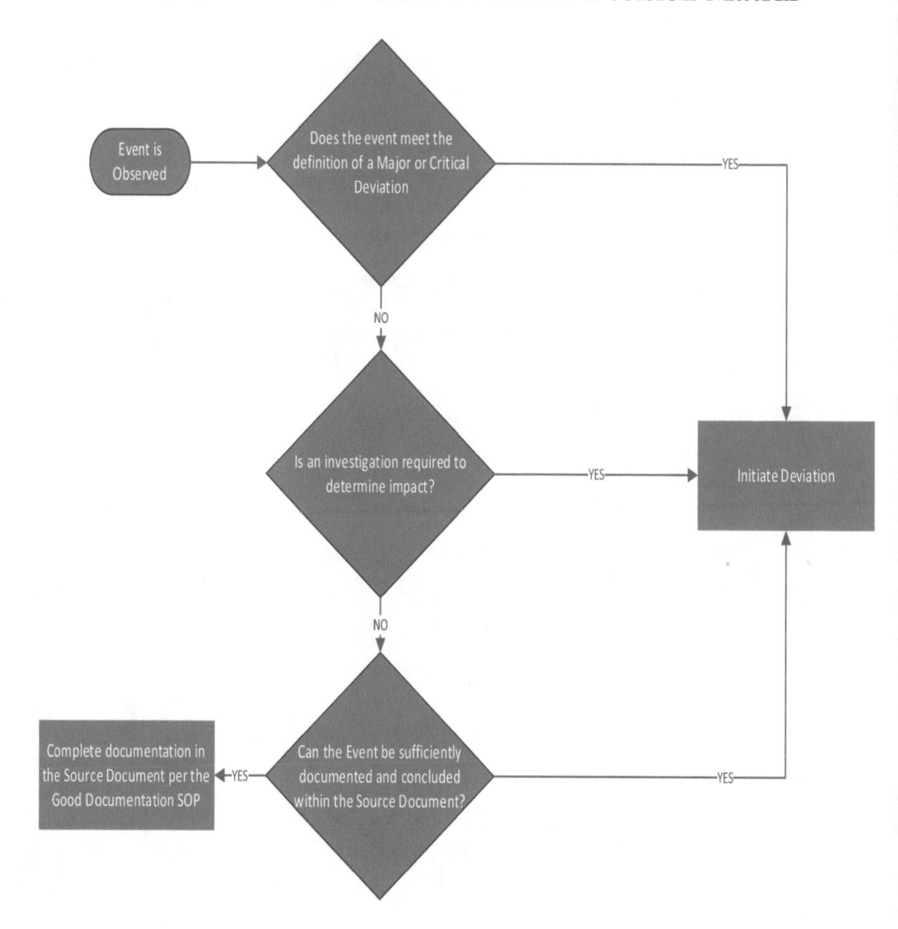

Attachment 3: Management Escalation Decision Matrix

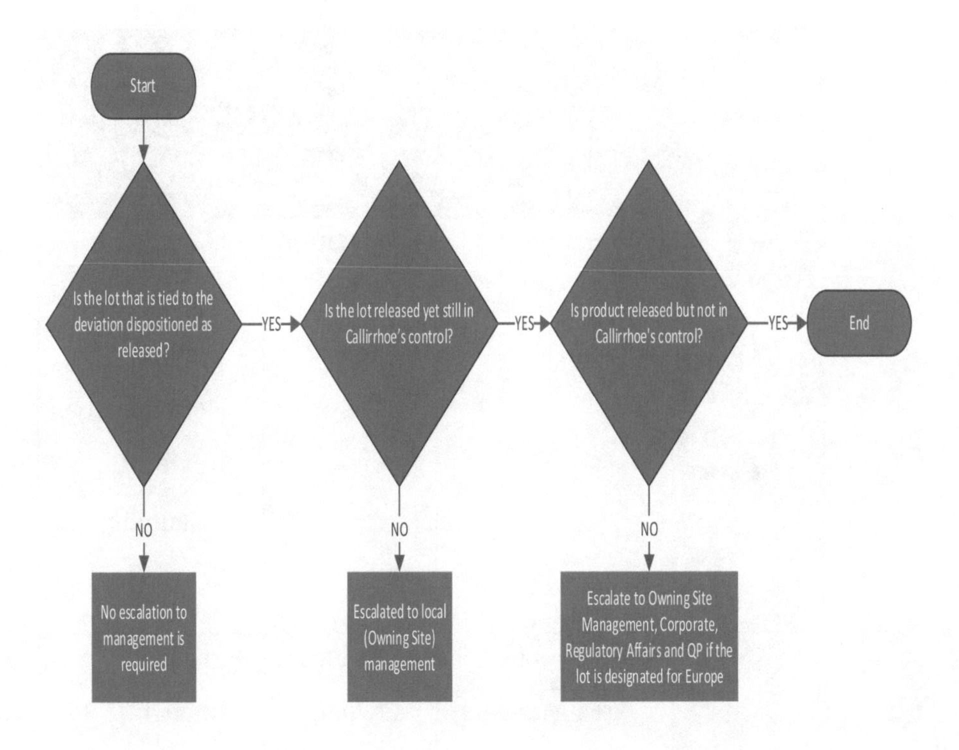

Attachment 4: Risk Based Approach to Managing the Deviating Event

Risk Category	Required Notifications	Required Approvers	Risk Risk-Based Approach
Minor	• Functional Area Manager • Quality	• Deviation Owner • Quality	• Document Root Cause • Perform Retrospective Review • Document Corrections
Major	• Functional Area Manager • Quality • Qualified Person (Europe batches only)	• Deviation Owner • Quality	• Product Impact Assessment • Impact Assessment • Perform Root Cause Analysis • CA for any identified Root Causes • Material status

Risk Category	Required Notifications	Required Approvers	Risk Risk-Based Approach
			evaluation & change if needed. • Investigation Summary / Conclusions.
Critical	• Functional Area Manager • Quality Director • Site Head Manufacturing Operations • Head of Quality (except for APH OOSs) • Qualified Person (Europe batches only)	• Deviation Owner • Quality • Head of Quality (except for APH OOSs)	• Product Impact Assessment • Impact Assessment • Perform Root Cause Analysis • CAPA for any identified Root Causes • Material status evaluation & change if needed.

Risk Category	Required Notifications	Required Approvers	Risk Risk-Based Approach
			• Investigation Summary / Conclusions. • PA's are required except for APH OOSs.

Appendix 2
Deviation Record Field Entry Guidance

1. Purpose

1.1. This work instruction provides guidance for field entries in deviation records and provides other general information for the use of (insert computer system, Veeva, Trackwise, Agile) as it relates to deviations.

2. References

1.2. N/A

3. Procedure – Deviation Record Field Entry Guidance

Data Field	Data Entry Guidance
Record Number	N/A – Computer system automatically creates the number.
Record Type	N/A – Upon initiation, the record is assigned by the Owner to the effective version of the deviation record.
Status	N/A – This field reflects the current status of the record as assigned by the computer system.
Title	Document a short description of the deviation: The short description is a brief statement describing the deviation. • Company acronyms and unique identifiers are to be defined. • For Trend Deviations, include the term "Trend" at the beginning. • Any short description or title for "planned deviation" is prohibited. • Ensure this fields is updated as necessary in order to be aligned with the deviation as new information is entered into the record.
Workflow	N/A – Represents the workflow the deviation owner assigned to the record based on the risk category.

Data Field	Data Entry Guidance
QA Approver	The QA Approver assigned to the record. The QA Approver cannot be the same individual as the Deviation Owner.
Owner	The Owner assigned to the record. The Deviation Owner cannot be the same individual as the QA Approver. • The Owner is the person responsible for coordinating the deviation documentation, investigation, and for the lifecyle of the deviation from Origination through Closure.
Date Originated	N/A – This date is assigned by the computer system when the record is initiated.
Date Closed	N/A – This date is assigned by the computer system when the record is closed.
Target Closure Date	The Target Closure Date should always reflect the current due date of the deviation record. Deviation record target closure dates are not eligible for extension.

Data Field	Data Entry Guidance
Initiation Site	From the List of Values, select the Site where the Deviation was initiated. • The Initiation Site, once selected, should not be revised at any point in the deviation record lifecycle. • If the Initiation Site is revised, enter a Key Event documenting the rationale for the change.
Owner's Site	From the list of values, select the Site that will Own the Deviation through the lifecyle of the record. • This selection is made based on where the deviation occurred, the area or site impacted by the deviation, or the Owning Site responsible for understanding the root cause, and implementing CAPA. • This field is critical to Inspection Readiness, Deviation Trending, and Work Center Teams or Management Review of Records. • If the Owner's Site requires revision, then update the field and enter a Key Event documenting the rationale for the change.
Deviation Owner's Department	From the list of values, select the Department that will be the responsible Department for Owning the Deviation through the lifecycle of the record. This selection is made based on where the deviation occurred, or area impacted by the deviation, and/or the responsible department for understanding root cause, or implementing CAPA.

Data Field	Data Entry Guidance
Deviation Owner's Department	• This field is critical to Inspection Readiness, Deviation Trending, and Work Center Teams or Management Review of Records. If the Owner's Department requires revision, then update the field and enter a Key Event documenting the rationale for the change.
Observed by:	Enter the name (First and Last Name) and title, or role, of the individual who observed the deviation. When there is more than one observer, enter the other observer(s) into the Description field.
Date Observed:	Date the event was observed. Observed Date is considered day zero. • This date may be different than the date of occurrence. • The observed date documented in the Description field must match the date of the Date Observed Field. • Future dates are not permitted.
Type:	High Level Categorization of the deviation. This field is critical to Deviation Trending. Only one Deviation Type is permitted per deviation record.

Data Field	Data Entry Guidance
Description	Provide a concise / full description of the deviation including: • What was deviated from? Describe what happened versus what was supposed to happen. • When it was observed and when it occurred. • Who was involved and who observed the event. • Extent – The size or magnitude of the event (How Much / How Many) • Context to understand the details or background of the deviation including the scope of product/materials involved, or equipment, facilities, and/or utilities. • Describe whether the deviation is Internal or External. • Do not include notifications, corrections, or containment actions, impact analysis, root cause analysis, or CAPA in this field.
Immediate Correction	Provide a concise / full description of any actions taken to eliminate, mitigate, or contain the detected deviation. Immediate Corrections/Containment Actions should include: • What: Short Description of the correction / containment • Who: First and Last Name and job title of who completed the action. • When: Date when the action was complete. • Result: Outcome of the action.

Data Field	Data Entry Guidance
Deviation Risk Category	Select minor, major, or critical based on the deviation risk definitions.
Rationale for Deviation Risk Category	Provide a robust justification detailing the rationale for the selection made in the Deviation Risk Category ensuring all elements of the selected definition are addressed. QA makes the final determination.
Products Impacted	Ensure that all product in scope of the deviation is connected to the Deviation by selecting the applicable Impacted Product Names. If the deviation does not have any product to connect, then select N/A.
Product or Material Lots	Enter the specific lot numbers to be tied to the deviation.
Impact Assessment Owner	The Deviation Owner or Functional Area Management determines the IA Owner. Insert the name of the owner into the field.

Data Field	Data Entry Guidance
Impact Assessment Target Due Date	Document the IA Target due date. Impact Assessment Due Dates will not be extended.
Impact Assessment	Product Impact Assessments are authored by Product Sciences. Product Sciences may elect to use a template.
Related MRBs	Enter all related MRBs. If not applicable, select N/A
Product Impact	Select "No" if there is no impact to product. Select "Yes" if there is impact to product.
Impacted Sites	Select all applicable sites that are impacted by the deviation.
Regulatory Escalation	Select the appropriate choice: • N/A – No Escalation Needed • Escalated – No Reporting Required • Escalated – Reporting Required

Data Field	Data Entry Guidance
Key Events (AKA Chronology of Events)	Document significant events during the documentation of a deviation: • Significant meetings such as MRB, or Regulatory Escalation. • Key Decisions (convening MRB, escalation, or material status changes). • Required notifications • Required justifications • Revisions to the initial quality approved deviation risk category. • Revisions to the Lot Tie Strategy or any changes in scope. • Changes in lots that are tied or removed. • Receipt of key information / test results. • Rationale for continued processing. • Justification for being past the target due date. • Multisite impact outcomes and mitigation strategy when applicable. • Transfer of Deviation Ownership or Transfer of QA Approver • Reopened deviations and notifications. • Use of the Business Continuity Process (BCP). • Site Quality Engineering Quality Risk Management Lead notifications for a Key Risk Determination. • External Deviation or Foreign Matter or Particulate Internal Deviation Quality Director or above due date adjustment approvals. Chronology of events when it helps to understand the deviation details or the

Data Field	Data Entry Guidance
	deviation impact analysis, or deviation root cause analysis. Key Events should be documented with clarifying details and full identifying information of personnel involved in the key event. Example: 29MAR2024. Head of Quality, Michael A. Delitala, Director of Quality Systems & Compliance and owner of the QMS, approved the due date adjustment to 90 calendar days from the date observed for the deviation in scope which is an internal particulate deviation. The date observed is documented as 01MAR2024. The target closure date will be adjusted to 30MAY2024. The particulate identified required outside vendor testing, review of the results internally to the company, interpretation of results, and final key decisions on how to process the lot tied to the deviation. The estimated completion is expected to occur by 15MAY2024.
Root Cause Analysis and Investigation Results	Refer to the RCA SOP.
Retrospective Review Results	Document: • The query parameters used to complete the retrospective review so that the query can be reproduced. • The decision for the time period used for the query. • The decision for relevance of any results obtained from the query.

Data Field	Data Entry Guidance
Root Cause	List the root cause(s)

Appendix 3
Training – Events versus Deviations

Evaluating Events: Event or Deviation?

What to Say...
Introduce yourself and welcome everyone to the class. Describe at a high level why this class is important and what we'll accomplish: • Today, we will go through training that will help us understand the differences between 1) an event and 2) a deviation. • We'll also describe why this is critical to get this right-first-time, and right, all the time.

What to Say…	Content
• Before we begin today's instruction, I'd like to get to know all of you. This helps me understand the varying levels of experience in today's class. • And based on the degree of experience, I may be able to adapt what and how is presented. **Note**: If there is an opportunity to engage the trainee based on their responses, then please do so.	• **Have an easel or whiteboard ready to capture answers from the trainee.** Session 1: Introductions 1. What Site do you work at? 2. What is your name and position? 3. What is, or what will be your role in the QMS? 4. What expectations do you have for this course?

Content

Session 2: Learning Objectives – Introducing Standard Terms

1. Define Source Document
2. Define Event
3. Define Deviation

Content

Standard Terminology – Source Document

Can anyone tell me what a **Source Document is?**
Per the deviation management SOP, a Source Document is:

- The origin of a documented entry such as but not limited to:
    - Manufacturing Production Records
    - Equipment Logbooks
    - QC Lab Notebooks
    - Analytical Test Methods
    - Facility Alarm Logs

The Source Document was the document being executed or documented on *when* or *where* an Event was observed.

Content

Standard Terminology – Event

Can anyone tell me what an **Event** is**?**
Per the deviation management SOP, a Event is:

- An unexpected or unplanned occurrence or situation during GxP operations that has the potential of being a deviation from:
    - cGMP
    - Regulatory Submissions
    - Validated Parameters
    - Standards, Policies, or Procedures
    - Specifications

We follow SOPs and execute a defined plan like a batch record or analytical method. Sometimes, things do not go as planned. That's an Event.

Content

Standard Terminology – Process Comment

Can anyone tell me what a **Process Comment** is**?**
Per the deviation management SOP, a Process Comment is:

- An entry or note annotated in GMP Source Documentation that describes an event.
- Process Comments are low or no risk events that can typically be immediately corrected and that do not require impact assessments nor root cause.

Process Comments are documented in Source Documentation to ensure events are documented and evaluated based on risk:

- **No or low risk events can remain an entries in source documents.**
- **Impact events, or events with risk, are elevated to a deviation.**

Content

Session 3: Learning Objectives – Evaluation Process

1. A note about the importance of Timeliness
2. Roles: Observer / Quality Assurance / Deviation Owner
3. Process Flow to Decide between Event or Deviation

Content

A note about the importance of Timeliness

- Timely decision making is critical for events and it is imperative that we get the decision right-first-time.
- That is why it is also important that the right staff are notified.

If you don't already know, take time to get to know your site's business of decision making.

It is critical that we decide swiftly on events versus deviations so that we ensure we're only sending quality product to our patients.

Content

There are Three Required Roles in the Event Evaluation Process

Observer

- Observes an Event and completes immediate corrections to protect product or personnel.
- Documents the Event at the Source Document.
- Notifies Functional Area Management and QA of the Event.
- Makes an initial determination about the Event being a Process Comment or a Deviation.

Quality Assurance

- Reviews the documented Event and confirms the decision.
- Documents their review and confirmation in the Source Document.
- If deemed a Deviation, then QA will request that a deviation be initiated.
- QA also ensures the deviation number is referenced back to the entry in the Source Document.

Deviation Owner

- Observes an Event and completes immediate corrections to protect product or personnel.
- Documents the Event at the Source Document.
- Notifies Functional Area Management and QA of the Event.
- Makes an initial determination about the Event being a Process Comment or a Deviation.

Functional Area Management also plays a significant role in this process ensuring personnel, product, and patients are protected through any additional immediate actions not initially completed during the Event.

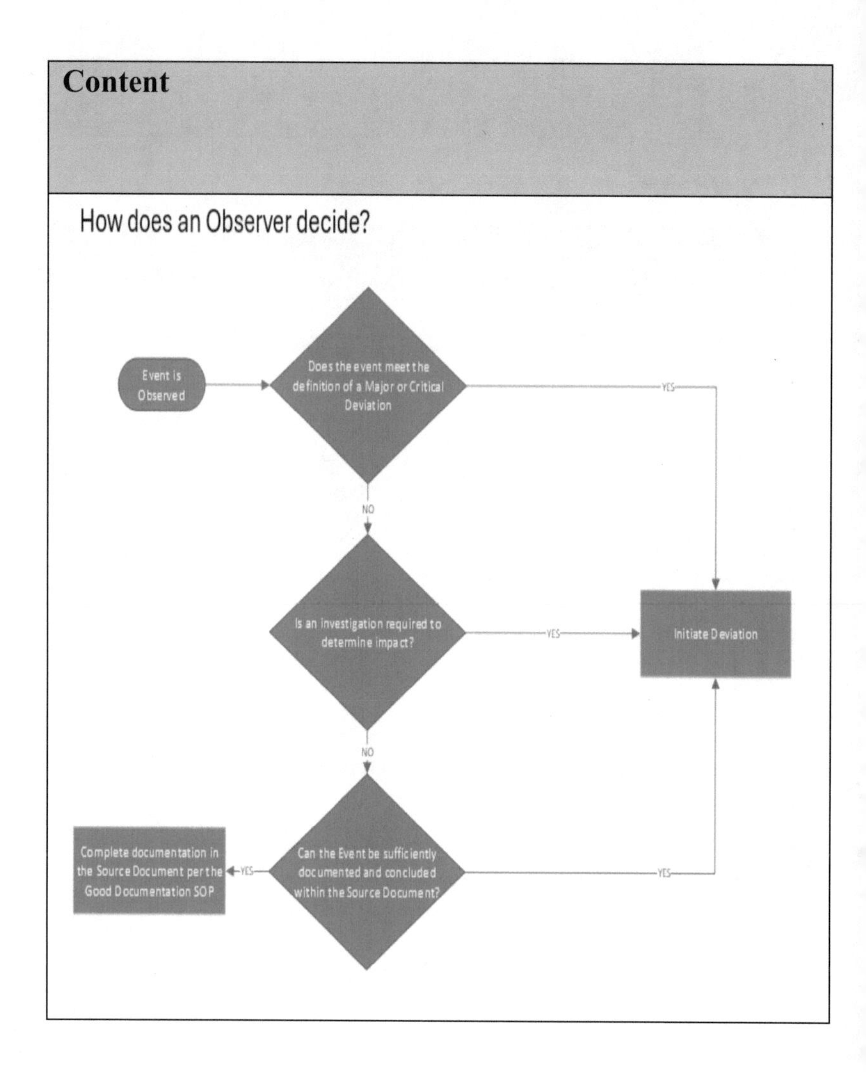
Content
How does an Observer decide?
Event is Observed
Does the event meet the definition of a Major or Critical Deviation
YES
NO
Is an investigation required to determine impact?
YES
Initiate Deviation
NO
Can the Event be sufficiently documented and concluded within the Source Document?
YES
YES
Complete documentation in the Source Document per the Good Documentation SOP

Appendix 4
Training – Minor Deviations

What to Say...
Introduce yourself and welcome everyone to the class. Describe at a high level why this class is important and what we'll accomplish: • Today, we will go through training that will help us understand how to think about and manage minor deviation. • We'll also describe why this is critical to get this right-first-time, and right, all the time.

What to Say…	**Content**
• Before we begin today's instruction, I'd like to get to know all of you. This helps me understand the varying levels of experience in today's class. • And based on the degree of experience, I may be able to adapt what and how is presented. **Note**: If there is an opportunity to engage the trainee based on their responses, then please do so.	• **Have an easel or whiteboard ready to capture answers from the trainee.** Session 1: Introductions 1. What Site do you work at? 2. What is your name and position? 3. What is, or what will be your role in the QMS? 4. What expectations do you have for this course?

Content

Your role in the deviation process is critical!

As a Deviation Owner, you are communicating important information to regulators, auditors, peers, and management.

- **The deviation record communicates the event and demonstrates how the company controlled a situation. A deviating situation.**
- **Your "audience" may include people who have English as a second or third language.**

Timing is key.

- **As the Deviation Owner, you should seek the information needed and generate the deviation record as timely as possible.**
- **Best practice is to raise and close the minor deviation within 5 business days from the initiated date.**

Those who do well with deviations often times find themselves being promoted.

Content

What is a *Deviation?*

- A departure from a written and approved procedure or other established standard for GMP operation and/or an unexpected event or failure.

- Deviations are categorized by their risk level: Minor, Major, and Critical.

Content

Deviations are classified by risk level: Minor. Major. Critical.

- Minor: No or unlikely potential risk.
- Major: Moderate potential risk.
- Critical: Significant or actual risk.

Content

Deviations are classified by risk level: Minor. Major. Critical.

- Minor: No or unlikely potential risk.
- Major: Moderate potential risk.
- Critical: Significant or actual risk.

The Deviation Owner assigns the initial risk classification based on the classification definitions.

The Quality Approver reviews the details and decides on the final classification.

Content

Define the Deviation Classifications

Minor	Major	Critical
Deviation with no potential impact or negligible risk to product quality, process, safety, efficacy, data integrity, and/or systems. No impact to CPPs or CQAs.	Deviation with potential moderate impact to product quality, safety, efficacy, data integrity, systems, and/or regulatory filing. Potential impact to CPPs or CQAs. Determination of Product Impact requires an investigation and analysis of routine testing or existing data. Additional routine tests may be required.	A deviation with confirmed, or high likelihood of significant impact to product quality, safety, efficacy, data integrity, systems, and/or regulatory filing. Impact to Critical Processing Parameters (CPPs), Critical Quality Attributes (CQAs) and/or product impact is evident or determination of product impact requires additional non-routine tests or extensive analysis of existing data sets.

Content

Best Practices to Arrive to the Correct Classification

1. Start with the description by reviewing what was deviated from.
2. The description provides the **context** by which the event can be compared against the definitions.
3. Consideration must be given to the impact on the product, process, and system.
4. Refer to examples already in the system by performing a query using similar terms.

Content

Best Practices for a Minor Deviation Short Description

1. A short description is a brief statement describing the deviation at the highest of all levels.
2. It is used for trending and historical searches.
3. For Trends, include the word "Trend" at the beginning of the short description.
4. For event notifications to suppliers include the abbreviation "EN" at the beginning of the short description.
5. Update the short description as applicable to align with deviation record.

Content

Write a Full Description of the Deviating Event

WHO	• Who was involved in the deviation? • Who observed it? • Who was notified of the event?
WHAT	• What deviated? • What acceptance criteria or parameter has been deviated from? • What was expected to happen and what actually happened? • What equipment, systems, products, or procedures were involved?
WHEN	• When was the deviation observed and noted? • When did the deviation occur? • Were there specific start and stop times?
WHERE	• The location where the deviation occurred. • The location where the deviation was first observed.
HOW MUCH / MANY	• Quantify the event (if applicable)
HOW OFTEN	• Frequence of the event (if applicable).

This is known as the 4W2H method.
The 4Ws are always required.
The 2Hs are optional.

Content

Next Step (now that you have classified the deviation), **Provide Justification for the Classification**

Minor
Deviation with no potential impact or negligible risk to product quality, process, safety, efficacy, data integrity, and/or systems. No impact to CPPs or CQAs.

Determination of no impact to product quality is possible with minimal review of data available at the time the deviating event was observed.

Documents available to assist with the determination include (not limited to):

- **Training records**
- **LIMS test results**
- **Work Orders**
- **Assessments approved for other events**
- **Real-time interviews**

Content

Best Practices when documenting the Classification

1. Classification should occur as an outcome of the event triage which occurs as close as possible to real time.
2. **NOTE: Minor deviations do not require an investigation, therefore it is essential that claims and rationales adequately justify no impact to product or process.**
3. Any supporting documentation (technical reports, data, summaries of data) should be attached to the record when not retrievable from another GxP source.

Bonus Content: Technical Writing – Claims, Reasons, & Evidence

Best Practices for Technical Writing – Claims, Reasons, and Evidence

1. Write clear and concise statement.
2. Base arguments on facts.
3. Describe relevant elements of the event so that someone who was not present can get a clear picture of how the elements (how the description) supports the classification.
4. Be detailed, yet, avoid superfluous information.
5. SOP, FORM, etc must be listed along with the document revision number all times it is applicable.

Bonus Content: Technical Writing – Claims, Reasons, & Evidence

Principles of Technical Writing: Claims/Reasons/Evidence

Principle	Definition
Use bottom line on time (BLOT)	• Put the main point first and follow up with the details. • Essentially, this is the CLAIM.
Apply claim, reasons, evidence pattern for main points.	• **Claim:** A sentence asserting something that may be true or false and because it could be both, it needs support. • **Reasons:** Statements that give readers cause to accept the claim. • **Evidence:** Facts or data that support reasons. There is no product impact. → Reason why there is no impact → Evidence for no impact
Use "chunking" to organize content	Organize related ideas together.

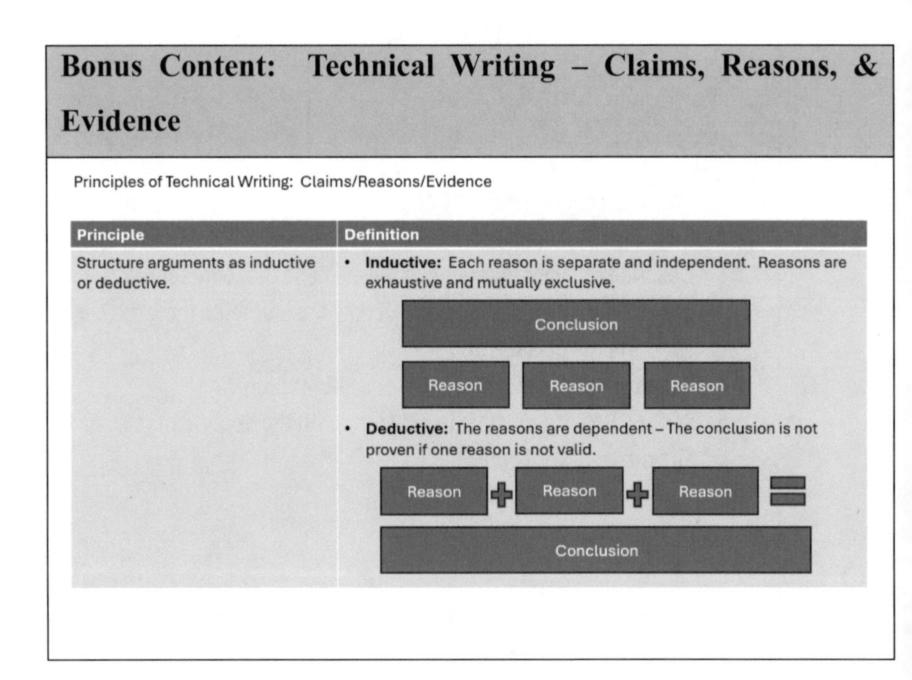

Bonus Content: Technical Writing – Claims, Reasons, & Evidence

Principles of Technical Writing: Claims/Reasons/Evidence

Principle	Definition
Structure arguments as inductive or deductive.	• **Inductive:** Each reason is separate and independent. Reasons are exhaustive and mutually exclusive. Conclusion Reason Reason Reason • **Deductive:** The reasons are dependent – The conclusion is not proven if one reason is not valid. Reason + Reason + Reason = Conclusion

Dear Reader, that certainly does not conclude the training. There are examples and many other nuggets but I simply cannot give it all away. For more, contact me through my website or my LinkedIn page.

Appendix 5
Bonus Material: Pre-Pivotal Event Evaluation Process

1. Purpose & Scope

1.1. The purpose of this standard operating procedure (SOP) is to provide instructions for documenting and assessing production and laboratory events during GMP manufacturing and testing of pre-pivotal materials.

1.2. Additionally, certain laboratory events, which may occur during pre-pivotal stability testing, as outlined in SOP-12345, may follow the Laboratory Event section of this SOP.

2. Responsibilities

Role	Responsibilities
Production or Laboratory Personnel	• Documents the event in the source document as close to real time of event occurrence as possible (i.e., timely manner). • Notifies the relevant Technical Function of an event. • Participates in evaluating the event with the relevant technical function as applicable. • Participates in evaluating the event with the relevant technical function as applicable. • Initiates a deviation record when applicable.
Technical Function	• Participates in the event evaluation process. • Provides definitive recommendation on the impact level when evaluating impact to product quality or patient safety. • Decides when an event is a deviation. • Initiates a deviation record when applicable.
Quality	• Participates in the event evaluation process. • Reviews events during batch record review or disposition processes. • Partners with the Technical Function to provide advice on events and event outcomes.

3. References

3.1. SOP-12345

3.2. SOP-23456

4. Definitions

Term	Definition
Certificate of Analysis (CoA)	An approved record for a given batch containing analytical test results required by specifications for the product or material.
Source Document	The firsthand, primary reference where the event was recorded. The source document is typically a batch sheet (BS) or manufacturing procedure (MP) but could be other GMP sources such as a notebook.
Event	1. Departure from approved instruction or defined standard. 2. Unexpected outcome or situation or an observation that is not part of the planned production process, not part of the planned testing process or method, or not defined in an effective standard operating procedure.
Routine Testing	• A standard catalogue of procedures or methods that are regularly performed. • Routine testing procedures are intended to demonstrate process performance during manufacturing processes or from representative manufacturing process samples, and are completed to demonstrate acceptance of a manufactured lot against a production parameter (in-process tests) or a specification (release tests)

Term	Definition
Non-Routine Testing	• Procedures or methods that may not be standard or not regularly performed. • Non-routine testing might be a part of the area catalogue of available tests but would be considered experimental for purposes of the batch in production or on test. • Non-routine Testing are tests designed explicitly for additional product or process understanding used to acquire additional data or demonstrate product quality or patient safety beyond in-process or release testing.
Root Cause	The underlying reason for the event.
Correction	Immediate action performed to eliminate or mitigate the impact of the event.
Technical Function	Laboratory or Production personnel, Functional Area Lead, or Functional Area Management
Pre-Pivotal	Stage of commercialization that begins prior to initiation of Phase 1 clinical trials and goes through Phase 2 clinical trials.
Phase 1	Initial introduction of an investigational new drug into humans.
Phase 2	Controlled clinical studies conducted to evaluate the effectiveness of the drug for a particular indication or indications in patients with the disease or condition under study and to determine the common short term side effects and risks associated to the drug.

5. Equipment & Materials – N/A

6. Safety – N/A

7. Procedure

7.1. General Considerations

7.1.1. The Technical Function has the highest level of product or process expertise in pre-pivotal processes. Therefore, the Technical Function is accountable for evaluating and deciding the impact of an event and decides whether an event will be documented in a Source Document or as a Deviation. This allows for flexibility in the quality management system while maintaining the necessary control to ensure product quality and patient safety.

7.1.2. Quality Assurance reviews events during batch record review and disposition processes. Quality Assurance can also act as an advisor to the Technical Function for events when applicable.

7.1.3. Events are expected during production or testing of pre-pivotal materials since pre-pivotal processes may be less understood or not fully characterized and not validated. Corrections or Corrective Action Preventive Action is inherent to event resolution.

7.2. Production & Laboratory Events

7.2.1. Document the Event

7.2.1.1. When an event is observed, document the event in the Source Document as timely as possible ensuring adherence to Good Documentation Practices per SOP-23456.

7.2.1.1.1. If the event is documented outside of the production record, then ensure the Source Document is attached to the production record.

7.2.1.1.2. Alternatively, if the event was documented in a GMP system, then document a reference with the production record.

7.2.1.1.3. Events may be documented in a memorandum to the product record if space is unavailable. Alternatively, the Technical Function may create a standard form for event documentation so long as it meets the requirements set forth in this SOP.

7.2.1.2. At minimum, document the following:

7.2.1.2.1. **Description:** Document a complete and concise description of the event details: Who, Where, When, and What. As applicable document How Much or How Many.

7.2.1.2.2. **Product Impact Assessment:** Document a description for product impact indicating there is or there is no impact to product. Include a rationale to support the assessment claim.

7.2.1.2.3. Corrections: Document immediate actions enacted to correct the event, if any.

7.2.1.2.4. Root Cause: If applicable, or if the root cause is known at the time of the event, then document the root cause.

7.3. Evaluate the Impact of the Event

7.3.1. The relevant Technical Function evaluates the details of the event in real-time, prior to Technical Function production record review, or during Technical Function production record review to assess the potential impact to product quality or patient safety.

7.3.2. The Technical Function determines an impact outcome which further decides how the event will be ultimately documented:

7.3.2.1. As an entry, memo, or functional area form to the production record.

7.3.2.2. As a Deviation.

7.3.3. Evaluation Leading to No Impact

7.3.3.1. If the event is evaluated as no impact to product quality or patient safety, then the Technical Function may accept the event as documented.

7.3.4. Evaluation Leading to Impact

7.3.4.1. If the event is evaluated to have impact to product quality or patient safety, then the Technical Function will initiate a deviation per SOP-34567.

7.3.4.2. IF the event impact evaluation is unknown or undetermined and requires:

- Non-routine testing or experiments to acquire data for additional product or process understanding to demonstrate product quality or patient safety beyond in-process or release testing OR
- A root cause investigation OR
- Additional data that will not become available prior to initiating a QA production record review

THEN the Technical Function will initiate a deviation record.

7.3.5. All event documentation must be complete and attached or referenced to the production record prior to the Technical Function production record closure and prior to initiating the QA production record review.

7.4. Quality Assurance Review of Events

7.4.1. Quality Assurance reviews events within Source Documents during the preview of production records per SOP-45678.

8. Attachments – N/A

Appendix 6
Deviation Types

- Deviation Types are high level categorization of the deviating event and is intended to aid in a Deviation Trending Program. Meta-data is power, and Deviation Types, when listed correctly, give anyone great power in trending deviation data.
- The Deviation Type is based on the deviation failure mode, deviation short description, or causal factor because the root cause is tracked separately.
- Deviation Types are not root causes.
- The Deviation Type is selected based on available information at deviation initiation. Ensure the type is correct throughout the lifecycle of the deviation record.

Deviation Type	Usage	Examples
Annual Product Review (APR)	APR process capability below acceptable level.	• APR identifies analytical results or process data with process performance index below acceptable level.
Acceptance Quality Level (AQL) Failure	Applicable to failures related to a sampling plan (includes but not limited to visual, manual, automated, and incoming inspections)	• Minor or cosmetic defects exceed established criteria. • Acceptable/reject limits exceeded for particles. • Semi-Automatic Testing Machine failures for a deliverable volume.
Complaint	Use for Customer Complaint Investigations	• Confirmed packaging or labeling errors. • Confirmed product specification failures.
Computer System	Use for any deviation that is related to software or hardware.	• Unplanned system outages. • Automation did not perform as expected. • Data migration errors.

Deviation Type	Usage	Examples
Confirmed Out of Specification (OOS)	Analytical test for product, process, and stability OOS.	• Valid OOS • Valid Stability OOS
Confirmed Unexpected Analytical Result (CUAR)	CUARs	• Valid CUARs • Valid Stability CUARs • Confirmed Lab Out of Tolerance (OOT)
Contamination	• Mircrobial, viral, or chemical contamination of a batch. • Presence of impurities or foreign matter that may impact the purity or utility of a batch.	• Mycoplasma • Bioreactor • Particulate matter • Residue observed on equipment.

Deviation Type	Usage	Examples
Damaged Material / Product	Physical flaw / damage in material or product during receipt, use, storage, or distribution	• Visually soiled gowns • Non-conforming post release materials requiring investigation to determine if suitable for use. • Nonconforming post release materials that are damaged. • Any batch that presents as damaged post-release and still in company's control. • Pre-use filter integrity failure.
Document Incorrect / Missing	Deviations where a document is missing, does not exist, or does not include instructions.	• Incorrect or incomplete instructions. • Instructions are not approved or are unavailable. • Change Notification received from a supplier when the change is closed. (AKA, Late Change Notification).

Deviation Type	Usage	Examples
Environmental Monitoring	Action limits for EM are exceeded.	• Gowning excursions. • Surface samples, total air particulate, air viable EM, and settle plate action limits exceeded.
Equipment Failure	The instrument or device physically fails or does not pass calibration and/or is unable to perform as designed.	• Inoperative balance. • Does not turn on. Broken. • Calibration failure. • Unlisted interventions.
Lost / Missing Batches	Batches are missing.	• Attempted or confirmed theft of batches. • Batches en route or in transit lost.

Deviation Type	Usage	Examples
Operational Parameter Excursion	Deviating event where an operational parameter does not meet the expected outcome as defined in a GxP document.	• Column equilibration did not meet specified pH range as required by the batch record. • Defects rate exceeded.
Pest Control	Deviating events with pests.	• Insect identified in BSC. • Spider hanging on a pipette. • Live insect on my glove.

Deviation Type	Usage	Examples
Planning / Scheduling	Action or time bound activities delayed or not performed by target date.	• Batch aborted due to business decision. • Missing calibration interval. • Missed stability time point. • Not enough time to produce the item. • Notification not sent on time.
Performance Parameter Excursion	A deviating event by which an in-process performance parameter has not been met.	• Raw material sample size does not match raw material specification. • Final day viable cell density excursion. • Shipper exceeded qualified time during distribution.

Deviation Type	Usage	Examples
Task Performance	Deviations from approved procedures such as: • Using the incorrect controlled document. • Transcription errors that have impacted data integrity. • Incomplete task • Reconciliation	• Cycle count discrepancies. • Miscount of shippers. • Task performed out of sequence. • Line clearance discrepancies. • Logbook entries do not reconcile. • Outdated controlled document used.
Training Requirements Not Met	Required training is not completed nor assigned.	• Staff executed a task without having prior training for the task. • Training had not been assigned.
Trend	QbD V3	QbD V3

Deviation Type	Usage	Examples
Unconfirmed OOS	A suspect OOS	• Analytical Testing area confirmed an unexpected lab result is an invalid result related to equipment or system.
Unconfirmed Unexpected Analytical Result	Unconfirmed Unexpected Analytical Result	• Analytical Testing area confirmed an unexpected lab result is an invalid result related to OOT results.

Made in the USA
Columbia, SC
03 November 2024